MARMOSETS

IN

CAPTIVITY

by Mike Moore

Basset Publications
Plymouth

Acknowledgements

Front cover - Painted by Stewart Muir of Shaldon Wildlife Trust, Reproduced by kind permission of Mrs. June Smith, Swindon.

Photographs of Marmosets - Reproduced by kind permission of Kilverstone Wildlife Park, Thetford, Norfolk.

Photograph of Cage - Reproduced by kind permission of Chris Morris, Plympton, Devon.

Line Drawings - By Stewart Muir and the author.

I dedicate this book to Bob Baker, with thanks for his support and encouragement.

First Published July 1989

I.S.B.N. 0 946873 95 X

Published by Basset Publications, 60, North Hill, Plymouth, S. Devon.

Printed by The Western Litho Company of Plymouth, Heather House, Gibbon Lane, Plymouth, S. Devon.

CONTENTS

Chapter 1

INTRODUCTION

Marmosets, (*Genus Callithrix*) are the smallest of the New World primates, and are to be found in the Northern half of South America (Fig.1). Their habitat ranges from tropical rain forest to the forest patches of the Amazonian savanna. They live mainly in the smaller shrubs and twigs of the tropical forests, which are not generally inhabited by larger animals.

The family name *Callitrichidae* is purported to mean 'beautiful hair', and the name marmoset comes from the French marmouset, which means 'young monkey', and in some translations, 'grotesque figure'. They are typically squirrel size, ranging from 29 - 38 cms, (11 - 15 ins.), in the Pygmy Marmoset, up to 44 - 50 cms, (17 - 20 ins.) in the other species, including tail. They have forward facing eyes, long tails, modified claws rather than nails (Fig. 2) and heads that can rotate through 180 degrees in either direction. Their coats vary in colour from dark brown to silvery white and have a silky lustre, there is a spectacular range of patterns within the group.

The group normally comprises a dominant breeding pair, single breeding females, and two or more sets of off-

PYGMY MARMOSET

TRUE MARMOSETS

Fig. 1. Natural distribution of Marmosets

1

spring, sometimes totalling up to fifteen animals. Social relationships within the group are maintained by mutual grooming which occurs throughout the average day.

Distribution of the different species and sub-species are as follows:

TASSEL EAR MARMOSET (*Callithrix humeralifer*)

These are concentrated within the Brazilian Amazon between Rios Madeira and Tapajos. There are three sub-species:

> *Callithrix humeralifer humeralifer*
> *Callithrix humeralifer intermedius*
> *Callithrix humeralifer chrysoleuca*

BARE EAR MARMOSET (*Callithrix argentata*)

These range from the Brazilian Amazon into East Bolivia and North Paraguay. There are three sub-species:

> *Callithrix argentata argentata* (Silvery Marmoset)
> *Callithrix argentata melanura* (Black-tailed Marmoset)
> *Callithrix argentata leucippe* (White Marmoset)

COMMON MARMOSET (*Callithrix jacchus*)

There are five sub-species, which have previously been described as separate species. These are:

> *Callithrix jacchus jacchus* (The Common Marmoset or The Cotton-eared Marmoset). These are found within the states of North East Brazil and have been introduced into Rio de Janeiro.

> *Callithrix jacchus penicillata* (The Black-eared Marmoset or The Black-tufted Ear Marmoset). These are found in the South central states of Brazil.

> *Callithrix jacchus aurita* (Buffy-tufted Ear Marmoset or White-eared Marmoset) These range within the states of South East Brazil around Rio de Janeiro and Sao Paulo.

> *Callithrix jacchus geoffroyi* (Geoffroy's Tufted Ear Marmoset or Geoffroy's Marmoset). These are also found within the states of South East Brazil, concentrated mainly in Minas Gerais.

2

Callithrix jacchus flaviceps (Buffy Headed Marmoset). These are again found within the states of South East Brazil within Espirito Santo and Minas Gerais.

PYGMY MARMOSET (*Cebuella pygmaea*)

This is the smallest marmoset and is placed, taxonomically, in a separate genus, *Cebuella*. They live in the Upper Amazonia in Colombia, Peru, Ecuador, North Bolivia and Brazil.

The Pygmy Marmoset and the other larger species are unique as gum eaters. They gouge holes in the tree trunks with their large incisor teeth. Using the upper incisors as an anchor in the bark, they gouge upwards with the lower incisors, producing holes which are oval in shape. They then lap the exudates from the holes. Some favourite trees can be literally riddled with these holes and up to one hundred individual holes have been observed in one tree alone. These exudates are a very large part of the Pygmy Marmoset's diet, but they are only a supplement to the other species. This can be important when other foodstuffs such as fruit may be in short supply. Some marmosets obtain these exudates from damaged trees, possibly those that have been attacked by burrowing insects. The Pygmy Marmoset spends nearly three quarters of its feeding activity tree gouging.

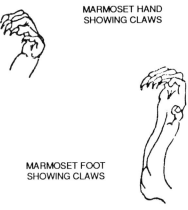

MARMOSET HAND SHOWING CLAWS

MARMOSET FOOT SHOWING CLAWS

Fig. 2. Marmoset hand and foot

Tamarins (*Genus Saguinus*), which can be confused with the marmoset species are also found within the Northern half of South America, chiefly inhabiting the tropical rain forest regions. In addition they are also found in secondary growth forests and semi-deciduous dry forests. Their size and weights are similar to those of the marmosets, and they also have fine silky coats enhanced by spectacular crests or manes in some species.

There are thirteen known species of tamarin, of which the Lion Tamarin is the largest, measuring some 78 cms. including tail. The tamarin diet is very similar to that of the marmosets, feeding primarily on fruit, saps and gums, flowers, tree buds, insects, small amphibians, such as frogs, lizards, and snails.

As already indicated both the tamarin and marmoset species look alike, but it is their dentition that separates them into two genera. (Fig. 3) The canine teeth of the Tamarin are larger, protruding well above the incisors, unlike the marmosets whose incisors are much larger. It is for this reason that the tamarin species have been classified as more dangerous and are included in the Dangerous Wild Animals Act, 1976, and therefore they are not readily available within the pet trade. The inclusion of the tamarin in the act came about in 1984.

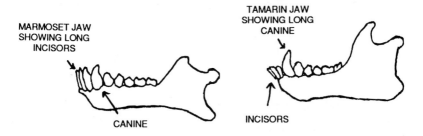

Fig. 3. Lower jaw of marmoset and tamarin

Although marmosets are very territorial, some tamarin species actually share a territory and the reasons for this have not as yet been discovered. The Saddle-back and Emperor tamarins are a typical example of this. It would seem possible that this behaviour must give certain benefits to both species, either defence of their chosen territories or possibly the detection of predators, as mutual calling between the different species occurs frequently, enabling the groups to maintain movement patterns. Their main predators are the forest hawks, and an animal similar to a weasel, but larger, called a Tayra (*Eira barbara*). Although the tamarin and marmoset species are adapted with long claws and needle like teeth, they have very little defence to a predator; instead they flee, taking refuge in holes in the trees, or in the

4

dense tangle of secondary growth. It is generally within secondary growth areas that most species will have their sleeping quarters because of the protection from predators that it offers.

Marmosets and tamarins both scent mark their territories. There may be up to ten groups or more holding territories within a one square mile area of primary and secondary forest. In addition to sleeping in the secondary forest most of the feeding occurs there too. Over a quarter of the activity time of marmosets and tamarins is taken up with foraging for food. This can be within the canopy of the trees, or shifting debris on the forest floor searching for insects, lizards, etc.

Tamarins and marmosets are usually described as monogamous, but field research has shown that adults of both sexes, mix within groups, although despite the presence of a number of females within a group usually only a single female is reproductively active. She may mate with more than one male. Care for offspring is predominantly by the male, but it is usual for all group members to take a share of the responsibilities.

Terborgh and Goldizen (1985) have recorded that over a three year period, groups contained one adult pair only 22% of the time, one adult female and more than one adult male 61% of the time, more than one adult of each sex 14% of the time, and one adult of each sex 14% of the time, and one adult male only 3% of the time. From this it is obvious that, in the wild, groups mainly contain more than one adult of each sex.

Chapter 2

BEHAVIOUR AND HANDLING

Marmosets are diurnal monkeys, active from dawn to dusk, with little naps in between activities. Resting can be within the security of the nest box, or on perches.

In captivity, as well as in the wild, marmosets walk, run, and climb on branches using a four-legged gait. They are efficient climbers, often stopping on their way to make some form of gesture, such as tongue flicking. Often they spring or leap from branch to branch and they may leap from vertical surfaces. They not only use the upper part of horizontal perches but also run underneath as well. Hanging by their back feet using their claws as an anchor is also a typical pattern of behaviour. Within the groups bouts of calling, chasing and displaying occur daily.

A typical display may show the rump, with the tail raised and fur fluffed up, and the white genitalia being shown. (Fig. 4) Mutual grooming takes place

Fig. 4. Typical display posture

often, normally between the adults, but also with and between, their offspring. This behaviour can be seen during most of the day, and is sometimes prolonged, possibly because they enjoy it.

Scent marking is also a common daily occurrence, usually being done with the chest and lower abdominal glands. Each and every branch and perch being covered in a strong smelling, sticky, secretion. In addition the glands around the genitals may be used; and often in the Silvery Marmoset the gouged holes in branches may be urinated into to communicate their dominance.

Foraging in the floor litter is also commonly seen throughout the day. This occurs naturally in the wild, when searching for insects and other small animals. In captivity crickets and other such insects can be introduced into the cage, either by simply scattering on the floor litter, or by using an insect dispenser which either allows the marmosets to take the prey as it emerges, or allows the insects to fall into the litter where they can be foraged for by the marmosets. For this reason it is important that hygiene standards are maintained at all times, although if deep litter such as bark strippings are used the risks of infection can be reduced.

Fig. 5. Arched back display

Outbreaks of aggression are unpredictable, although they often occur when the female is in oestrous. One sign of tension is the arched back display. (Fig. 5) As the female nears oestrous males will increase the frequency of associated behaviour such as grooming and huddling. Females show an increase in associated behaviour, directed towards the males, two to four days before the males' peak of sexual interest. (Kleiman 1977).

Although very agile and comical in their activities, marmosets are very sensitive animals and subject to stress. This is normally seen at its most extreme when the animals are moved to new enclosures, or locations. The fact that they have to be caught, then transferred into a completely fresh environment creates confusion and the animal can harm itself in the associated panic. When moving marmosets to new cages I have found that one or two perches from the old cage, put into the new cage, can be advantageous, as the familiar scent markings can help to settle the animals. Stress can also induce these animals into biting the tips of their tails, although this may not necessarily be caused by moving the animals.

When the whole family is to be moved, say to another cage, the movement should take place at night when the animals have settled down in their nesting box. The entrance hole to the nesting box is closed and the box simply transferred to the new cage. This dramatically reduces the stress put on the marmosets. It is only when one individual from the group needs to be caught, say for treatment that the problems are intensified.

Marmosets do not like to be handled, and if it is not done correctly, it can lead to lacerations from their sharp claws and severe bites to the hands. Gloves should be worn when handling a marmoset to protect against injury, but the catching of the individual still has to be done carefully. If the animals can be encouraged into the outside run and then allowed back inside, one at a time, by using the intermediary slide, (see Housing), the marmoset will eventually be caught. This operation may have to be carried out more than once though.

When the animal is separated from the group it can either be caught in a hand net, or shut into the nesting box. I have found that cloth nets are the best to use, as they eliminate the risk of the animals' claws catching in the mesh of an open mesh net. Once the animal is in the net, turn the handle to seal the opening of the net, and carefully feel for the back of the neck of the animal. By using gentle pressure either side of the neck using the first and second fingers, it is possible to restrain the animal. The back of the animal can then be cradled in the palm of the hand, still using the net as protection. Then slowly turn the handle so that the net opens, and the animal is exposed. This allows the underparts of the animal, its head, legs, feet and tail to be observed. To be able to inspect the back transfer the animal into the other hand; which should be gloved. This whole operation can be carried out in the cage, so that if the animal escapes from the net it is still confined.

When trying to net a marmoset, the animal will throw itself around the cage, so it is important to stand in one position if possible, and allow the animal to settle. Never use a net which has no padding around the ring, as bruising to the body can occur, and it has also been known for an animal to be concussed and even killed.

Chapter 3

HOUSING

It is important to construct suitable housing for marmosets, as they are hyperactive animals which run, climb, leap and forage for most of the day. Enclosures must contain suitable furniture and substrates to enable the animals to carry out all these activities. In the wild these animals live in primary and secondary forest, with leaves and other vegetation littering the forest floor. As far as possible this should be copied in a practical way by using natural perches, twiggy foliage, and for the floor, sawdust, woodshavings, peat, bark strippings, bark mulch, or a similar substrate.

Marmosets are not solitary animals, and need to be housed in pairs, when first starting to keep them; the cage ideally though should be big enough to house a family group. In my view if they are kept in too small an enclosure the animals daily activities are limited and this will be detrimental to their general well-being. If the animals are to be kept primarily indoors access to an outside enclosure should be made available, conversely if the animals are to be kept outdoors a suitable heated inside enclosure must be available to them at all times as they cannot suffer cold for long periods. Whether housed indoors, or out of doors a heated nest box is necessary as well.

When keeping the animals indoors I would recommend that the cage is a minimum of 1.8 metres (6 feet) high, 1.2 metres (4 feet) wide, and 1.2 metres (4 feet) deep. Obviously if a larger space is available then make the cage as big as possible. Once the floor litter is put down, and the heated nest box and the cage furniture installed then the total area available for the animals will be reduced. The indoor cage should be constructed on a solid base avoiding if possible the use of wooden floors, unless covered, as these will encourage dampness and could eventually start to rot and smell, possibly harbouring infection in the process. A concrete base is suitable if it is to be covered with deep litter. For the purposes of this book I would recommend the use of bark strippings as a deep litter as cleaning of the cage can be minimised, although waste food must be taken out daily. This litter also gives the animals the opportunity to forage, and absorbs most moisture.

The walls of the indoor cage can be made of heavy duty plywood, if a natural wall is not to be used. The wood must not be treated, but it can be painted with a non-toxic paint. As some marmosets do gouge wood with their lower incisor teeth, a brick built cage could be built, but this would be rather a cold environment, so the brick could be overlaid with plastic covered board or formica. This also makes cleaning easier if they are kept smooth. The front of the cage can be wired with a suitable wire mesh, 25mm x 25mm (1 ins. x 1 ins.) 16 gauge being the most suitable. An alternative is to glaze the front, bearing in mind the need for suitable ventilation. The ceiling of the cage should also be made from a rigid material, as this will help to hold the vertical poles of the cage furniture.

The furniture for the minimum sized cage described could be three upright stripped larch poles, about 5cms. (2 ins.) in diameter with a number of similar sized poles bolted to them to form horizontal runs. Once the main framework has been constructed twiggy branches can then be tied to it to create a more tree like appearance.

The heated nest box (Fig. 6) can either be included inside the cage or fitted onto the outside of the cage, with an intermediary slide to enable it to be used as a catch up facility. The box needs to be made from heavy duty plywood which will be left untreated, but could be painted with a non-toxic paint. The size should be 75cms. (30 ins.) high, 45cms. (18 ins.) wide and 45cms. (18 ins.) deep, with a hinged door 30cms. (12 ins.) high. The top 15cms. (6 ins.) of the box contains the heat lamp which is separated from the rest of the box by a mesh panel so that the animals cannot interfere with it. A 10 cms, (4 ins.) square entrance hole is cut in the door, 7.5 cms. (3 ins.) from the bottom. This gives access to the heated box for the animals, and is sufficiently large to cope with an adult carrying young. Within the main box is a smaller box, measuring 25cms. (10 ins.) high, 20cms. (8 ins.) wide and 20cms. (8ins.) deep. Again a 10cms. (4 ins.) square hole is made in the front of the box, but this time 5 cms. (2 ins.) from the top to allow the animals to go in, and down, for privacy. The animals can use the top of the small box for more direct heat, as it is

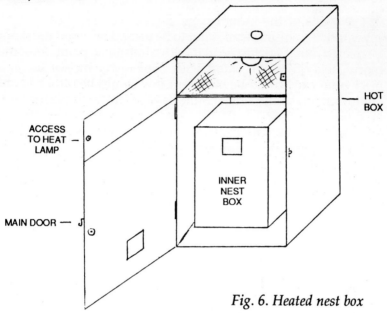

ACCESS TO HEAT LAMP

MAIN DOOR

HOT BOX

INNER NEST BOX

Fig. 6. Heated nest box

positioned directly below the heat bulb, which is protected by the mesh. I have found that this type of heated nest box gives the animals a choice of either very warm (75 °F, 24 °C), on top of the small box, adequately warm (70 °F, 21 °C) by being inside the small box or in the large box generally, or being at the ambient temperature in the fresh air of the main cage.

If an outside cage (Fig. 7) is to be used as the main activity area it needs to be the same minimum size as the indoor cage, but preferably larger. It can be constructed of a wooden framework with 16 gauge weldmesh attached. The floor can be the same as the internal cage, but if a natural earth floor is to be left then try to ensure that the ground is clean as infection from rodent and bird droppings can occur. A solution of Jeyes fluid and water can be applied to the soil and will kill most infections, (see the can for details of dilution rates). The roof should be part covered, to allow protection from the elements, and to protect against the risks of infection carried by wild bird droppings. Boarding at the bottom of the sides of the cage will prevent rodents from entering and contaminating the floor of the enclosure.

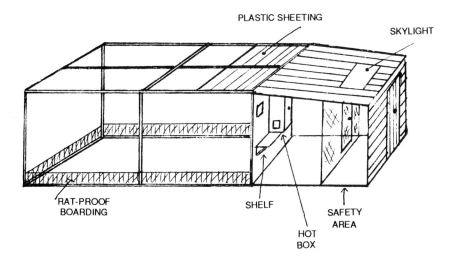

Fig. 7. Basic design for large outdoor cage before dressing with perches

Suitable bolts, top and bottom, are essential on all doors, and where possible a double door should be installed to prevent escapes. When installing electricity to the heated box, always ensure that the wire is protected or out of reach of the animals as they may bite through cables.

13

Chapter 4

FEEDING

As already outlined in the introduction, in the wild, marmosets feed on a variety of foodstuffs. Most of these foods can be supplied in captivity, although the exudates present a problem. At the Shaldon Wildlife Trust, a sweet nectar mix is prepared each day, to compensate for these exudates. This is made from infant food, milk powder, and sugar, mixed with warm water. Daily, fresh fruit is given with some vegetable matter, such as carrot; wholemeal bread and SA37 vitamin supplement. Live food such as crickets are given three to four times a week, and mealworms once a week.

A typical daily diet for a pair of marmosets would be:

$^1/_2$ banana
$^1/_2$ apple
$^1/_2$ pear
$^1/_2$ orange (peeled)
$^1/_2$ tomato
$^1/_2$ small carrot
1   dried fig
$^1/_4$ slice of wholemeal bread

All cut into half inch cubes and sprinkled with SA37 vitamin supplement.

The daily nectar ration is made from:

10g Milupa baby food (Fruit Flavours)
2g   milk powder
$^1/_4$ teaspoon sugar

This is mixed with warm water to form a creamy consistency.

The nectar mix can be given first thing in the morning, followed at mid-day with the main meal. Insects can be given during the day, or if an insect dispenser is installed, then the animals will cope with that themselves.

Other foods that may be tried are: day-old chick legs, pieces of cooked chicken, grapes, yoghurt, cream cheese, soft fruits, sunflower seeds, peanuts, tinned dog or cat food, root vegetables (cooked or raw), sweet biscuit, and hard boiled egg. Give in small quantities, to see what your animals prefer, but look upon these as additions to the normal diet.

There are many other diets that can be obtained, such as marmoset jelly. This is supplied as a white powder, with a powdered flavour that can be added. The powder is mixed with the flavouring and then dissolved in hot water and allowed to cool. It sets as a jelly which can be cut into cubes or strips and offered to the marmosets. This form of food is high in protein, and high in energy and gives the animals most of the nutrients they require. It can prove to be an expensive method of feeding though, and some marmosets do not take to it readily.

Marmosets require a high input of vitamin D_3 in their diet, and this is not obtained in the normal captive rations. A concentrate vitamin D_3 powder can be obtained from Roche Products Ltd., and this is administered onto the dry food after diluting it with glucose. Dilution is at a rate of 1 gram D_3 powder to 500 grams of glucose powder, being sure to mix them well. This mixture is sprinkled over the food at a rate of $1/_4$ of a teaspoon full daily. This should give an adequate level of vitamin D_3 and prevent any bone deformities from occurring. Other additives that can be given to marmosets to increase their vitamin intake are Cytacon B12, (which contains vitamin B_{12}) and Abidec (which contains vitamins A, B_1, B_2, B_6, C, and D_2), these are available from most chemists. A few drops weekly in the nectar mix is usually sufficient.

Water must be given daily, and all feeding utensils must be emptied, washed thoroughly and dried, each day before fresh food is put in them. Plastic 'D' cups can be used as food containers, but these eventually become scratched and can harbour germs. Stainless steel feed bowls are more suitable, and these can be obtained from most pet stores. A hooked ring fits onto the wire of the cage, and a round stainless steel bowl drops into the ring. Having no corners to trap food makes them easy to clean. Always make sure that the animals have easy access to the feed bowls by having a horizontal perch about 5 cms. from the bowl.

Chapter 5

BREEDING

Obtaining your first pair of marmosets can be a very exciting experience, but it is important to make sure that the pair come from different blood lines. If they happen to be related then breeding them could bring forth offspring which may be unhealthy, runted or deformed.

Sexing of marmosets can prove to be difficult, and the only obvious method is to examine the genitalia. (Fig.8).

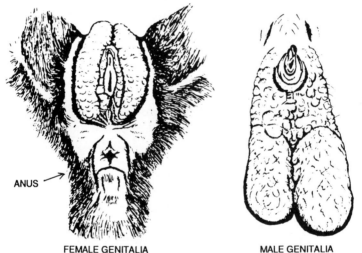

ANUS

FEMALE GENITALIA　　　　　MALE GENITALIA

Fig. 8. Female and male genitalia

Common Marmosets (*Callithrix jacchus*), normally breed quite readily, if the conditions are correct, from about eighteen months of age, and may produce two litters a year. Oestrus cycles in the female vary between 14 to 24 days. Most species have a post-partum oestrus, between 2 to 10 days after giving birth, at which conception can take place. The gestation period for the Common Marmoset is usually 140 - 149 days. Most litters are born either in the Spring or late Summer, but Common Marmosets will breed during most months of the year.

Births normally take place at night, or in the early hours of the morning. Observations of births show that the male stands behind the female and, once the baby has arrived, helps to clear and eat the afterbirth and places the baby on his back. When the second baby arrives the same procedure is followed. The male then carries the babies for the first few days, only passing them over to the female for feeding. After this period, both male and female take turns in carrying the young. Twins are normal, but triplets can occur. Triplets can present a problem as the female has only two nipples to feed the young. She will normally feed both youngsters at the same time, so a third youngster may be neglected and die. It is possible though to remove the young-ster for hand-rearing.

Hand-rearing is a very time consuming process, as for the first week the infant will have to be fed on a one to two hourly basis. Much successful hand rearing has been done in the past, and one of the most successful people in this field is Lady Fisher of Kilverstone Wildlife Park, in Thetford, Norfolk. Her suc-cesses now exceed one hundred individual animals. Methods of hand-rearing differ between marmoset owners, but at the Shaldon Wildlife Trust we have adopted a method similar to Lady Fisher's and are getting good results. It is this method that is described here.

To ensure that the infant is being kept at the right tempera-ture a plant propagator (set at about 25 - 28 °C, 77-88 °F) is an ideal incubator/nursery. A woolly jumper or soft towel should be put into the propagator, and something for the youngster to cling to, such as a small soft toy, will help to give it some security.

The smallest 'Catac' feeding bottle, available from most pet stores, is used with a ST1 teat. To regu-late the flow of milk, a teat, with-out a hole in it, from a human baby bottle is ob-tained and is

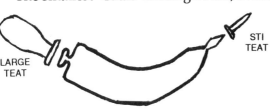

LARGE TEAT

STI TEAT

Fig. 9. Catac bottle (adapted)

placed on the end of the Catac bottle (Fig. 9). With a little pressure on this teat, it is possible to regulate the amount of milk leaving the bottle and entering the baby's mouth.

From my experiences with hand-rearing, the first five days are the most crucial. Although two hourly feeds are given, I have always offered extra food to the animal every hour. The animal may not take each additional feed, but this is the period when the internal organs need to receive as much nutrition as possible for proper growth. A young marmoset, in the Shaldon Wildlife Trust, which was being reared by its parents, died at five days old and at post-mortem was found to have under developed internal organs. This was attributed to the mother having insufficient nutrients in her milk.

After EACH feed, the abdomen should be wiped with a moist tissue to induce defaecation.

DAYS 1 - 5   S.M.A. is given at each two hourly feed. The dilution is one scoop added to 28ml. (1 fl.oz.) of boiled water, administered at body temperature. The youngster will normally take around 0.5ml. Into one feed daily put one drop of ABIDEC.

                Between two hourly feeds, give a glucose feed. One teaspoon of glucose powder to 28 ml. (1 fl. oz.) of boiled water. Around 0.5 ml will normally be taken by the youngster, at body heat. The remainder can be kept in the fridge, but always remember to warm the liquid to body heat before giving to the youngster.

DAYS 6 - 14   Continue with the SMA and ABIDEC as above, but also add a little Milupa infant food (fruit flavour). Mix the Milupa to a paste, and add to the milk. This is now giving extra bulk to the animal, and will help in giving some of the trace elements that it requires. In weeks two and three, start to introduce vitamin D_3 twice a week. This comes in powder form, and can be suspended in arachis oil, to make it easier to administer. The youngster will require 1000 i.u. per week to ensure that bone deformities do not occur.

WEEK 4   The feeds should now be around every three hours, and the SMA reduced and the Milupa increased. Continue with the ABIDEC and vitamin D_3. In addition the youngster may be offered a little mashed banana to introduce it to solid food. This will be a trial and error period, but do not try and force the animal to eat, it will do so in its own time.

WEEK 5   The youngster should now be taking some solid food, and other fruit can be added to the diet. Use mainly soft fruits and allow the animal to try a piece of carrot to chew.

The following weeks will show a sharp increase in the animal's diet, and several other foods can be introduced, (see chapter on Feeding). Bottle feeds should be continued as long as possible, to give a certain amount of security to the youngster, but these can be reduced as the animal more readily takes to feeding itself from a bowl.

At about nine weeks old the youngster can then be re-introduced to the adults. Rejection by the parents can occur, but if the parents are teased with the youngster, they will usually snatch it from the hand, and their protective instincts then take over.

Youngsters are left with their parents for at least fifteen months, before being taken away. This is to ensure that they have learnt, from the parents, how to care for offspring and other normal behaviour patterns. Inexperienced youngsters that have been taken away from their parents before this age, usually destroy their first litter, and possibly some later litters.

Chapter 6

BIOLOGY AND HEALTH

Biological Data:
Adult Weight:	250 - 450g.
Adult length:	head and body 18 - 30 cm.
	tail 17 - 40 cm.
Dentition:	32 teeth.
	4prs. incisors, 2prs. canines,
	6prs. premolars, 4prs. molars.
Females:	1 pr. of mammae

Longevity: An average lifespan of eight years has been recorded in captive animals with some attaining the age of twelve years. (Mallinson 1975). The longevity of marmosets in the wild is unknown, but as they are preyed upon by many species, such as forest hawks, it would seem that the lifespan would be dramatically reduced compared to that in the captive environment.

Health: Inadequate diet and bad housing can create disease and imperfections in the animals, but with a good variety of foodstuffs and reasonable heated housing the animals should stay in perfect health.

The most important supplement to the diet of marmosets is vitamin D_3. Because most captive marmosets are housed indoors, the absence of direct sunlight means that they are unable to make their own vitamin D_3 in their skin. This can cause rickets and osteomalacia in the animals. Each animal requires an intake of 1,000 i.u. of vitamin D_3 per week, and this can be simply administered in glucose powder, as per the chapter on feeding.

Illnesses, Prevention and Treatment: Draughts and dampness can be fatal to marmosets, and they are very prone to human ailments such as colds and herpes, as well as other respiratory viruses. So care should be taken to avoid passing these on to the animals.

Common Cold - A simple human cold can be a disaster to a marmoset. Pneumonia can soon develop and will prove to be fatal.

Dental Problems - Abscesses in marmosets are not uncommon, and are usually associated with the canine teeth.

Caries - a condition in which teeth decay and eventually disintegrate, in animals that are fed on man made sugary foods. Veterinary advice should be sought for any of these.

Diarrhoea - Loose stools can be passed by frightened marmosets, as a result of stress, or from too much fruit in the diet. Stools which are very fluid should be examined by the vet. If the animal continues to pass them it will lose weight and if untreated eventually die.

Escherichia coli - is a bacterial infection which can prove fatal to marmosets, and I have found that yearly vaccinations with 'Porcovac' have helped to eliminate the problem. Your vet should be able to help with obtaining and administering this vaccine.

Hepatitis - this virus can easily be transmitted from many species of monkey, including marmosets to man.

Herpes virus infections, including the common cold sore, can be transmitted to the animals from humans and may be fatal. Herpes virus B can be transmitted from marmosets to humans by a bite or scratch, and may cause serious illness. Most herpes viruses cause lesions on the lips, palate, cheeks and gums of the animal. Any animal suspected of being infected should be isolated from the others, using gloves or a net, (see handling and behaviour), and a veterinary surgeon called. The isolated animal should be kept warm (75 °F).

Measles - Can be transmitted from human children and adults, and may prove fatal to marmosets. Signs to look for are a rash, loss of weight, diarrhoea and discharge from the nostrils.

Mumps - Contracted from humans again, but affects young marmosets more than adults.

Pneumonia - can be passed on to marmosets by humans, the usual symptoms are, loss of appetite, loss of condition, respiratory distress and, if left untreated death.

Tuberculosis - causes emaciation, and if the lungs are involved, respiratory distress. Although the majority of infections arise within the group of monkeys, it can also be contracted from man.

All of these diseases have the potential to be very serious and some are communicable to man so strict hygiene should be followed with any sick marmoset and the veterinary surgeon called quickly.

Chapter 7

OBTAINING STOCK

I never recommend that people acquire marmosets as pets because it must be recognised that they are wild animals. They can make pleasant pets though, if properly housed and fed, but they are not animals that can be handled like a cat or dog; and they are not suitable to be free in the home or near children. Accordingly it is always advisable to obtain sufficient information on their care and husbandry before even considering buying them.

Common Marmosets (*Callithrix jacchus*), are obtainable in some pet stores, and they can be found for sale in Exchange and Mart as well as in the weekly paper Cage and Aviary Birds.

It is not advisable to obtain single animals, but to buy an unrelated sexed pair, which were at least fifteen months old before being removed from their respective parents. If taken away early the animals may not have learnt the natural behaviour patterns that are present in their parents, (see Breeding).

Always go to collect your animals, to ensure that they are healthy and active. They should be alert, with dense fur, bright eyes and no wetness around the anus. Sick animals will appear listless, with their eyes dull and fur ruffled.

Description:

The general body colour is grey, with the lower part of the back blackish and greyish, orange or tawny. This is normally in alternating bands, with the tawny or orange more or less concealed as an under fur. The tail has alternating blackish and pale bands. Long tufts of hair surround the ear area, and the forehead is white to nearly black.

Transporting the animals from the place of purchase should be done in a secure wooden box with ample ventilation. The lid can be hinged, with a 46 mm ($^1/_4$ ins.) mesh viewing panel about (6 ins. x 2 ins.). (Fig. 10)

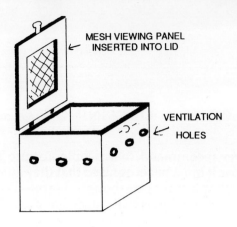

MESH VIEWING PANEL
INSERTED INTO LID

VENTILATION
HOLES

Fig. 10. Transportation box

When purchasing your animals, obtain a diet sheet from the seller and continue to use it. If another diet is to be introduced, do it slowly by adding a little to the established diet each day until the animals are fully weaned onto the new diet. If a new diet is introduced too quickly the animals motions may tend to become very loose.

Fisher, Lady, Hand-rearing marmosets and tamarins, from Kilverstone Wildlife Park, Thetford, Norfolk.

Hershkovitz, P. (1977) Living New World Monkeys (*Platyrrhini*) vol. 1 University of Chicago Press, London.

Ingram, J.C. (1975) Husbandry and observation methods of a breeding colony of marmosets (*Callithrix jacchus*) for behavioural research. Lab Animal **9**: 249 - 259.

Kleiman, D. (1977) The Biology and Conservation of the Callitrichidae. Smithsonian Institution Press, Washington, D.C.

Mallinson, J.J.C. (1975) Breeding marmosets in captivity. In, Breeding Endangered Species in Captivity (R.D. Martin, Ed.). Academic Press, London.

Sussman, R.W. and Garber, P.A (1987). A new interpretation of the social organisation and mating system of the Callitrichidae. Int. J. Prim. 8(4) 397

Sussman, R.W.and Kinzey, W.G. (1984). The ecological role of the Callitrichidae: a review. Am. J. Physical Anthropol., **64**, 419-449.

Terborgh, J. and Goldizen, A.W. (1985) On the mating systems of the co-operatively breeding Saddle-backed Tamarin (*Saguinus fuscicollis*). Behavioural Ecology. Sociobiol., **16**, 293 - 299.

A Manual of the Care and Treatment of Childrens' and Exotic Pets, edited by A.F. Cowie, British Small Animal Veterinary Association, 7, Mansfield Street, London W1M 0AT.

Dodo, The Scientific Journal of Jersey Wildlife Preservation Trust, Les Augres Manor, Jersey, C.I.

International Zoo Yearbooks, volumes 15, and 22, Edited by P.J.S. Olney, Published by the Zoological Society of London.

Management of Prosimians and New World Primates. Proceedings of Symposium 8 of The Association of British Wild Animal Keepers, Bristol.

The Welfare of Pet Marmosets, The Captive Care Working Party of the Primate Society of Great Britain, Published by Universities Federation for Animal Welfare, 8, Hamilton Close, South Mimms, Potters Bar, Herts. EN6 3QD.

*   ABIDEC multivitamin - Warner Lambert Health Centre, Mitchell House, Southampton Road, Eastleigh, Hants. SO5 5RY (0703 - 619791)

      CATAC feeding bottles - Catac Products Ltd., Catac House, 1 Newnham Street, Bedford. MK40 3JR. .(0234 60116)

*   CYTACON B12 - Duncan Flockhart and Co. Ltd., 700 Oldfield Lane North, Greenford, Middlesex UB6 0HD (01 - 422 - 2331)

      MARMOSET JELLY - Special Diets Services, P.O.Box 705, Witham, Essex, CM8 3AD, (0376 - 511260)

*   MILUPA Infant Food - Milupa Ltd., Milupa House, Uxbridge Road, Hillingdon, Uxbridge, Middlesex, UB10 9NA.

*+   SA37 Vitamin Powder - Intervet U.K. Ltd. Science Park, Milton Road, Cambridge, CB4 4FP

*   SMA Powdered Human Milk Substitute - John Wyeth and Brothers Ltd., Wyeth Laboratories, Huntercombe Lane South, Taplow, Maidenhead, Berks. SL6 0PH (0S286 - 4377)

+   VITAMIN D_3 - Roche Products Ltd., P.O. Box 8, Welwyn Garden City, Herts. AL7 3AY (0707 - 328128).

      * Available from pharmacists

      + Available from/through your veterinary surgeon